GW00371270

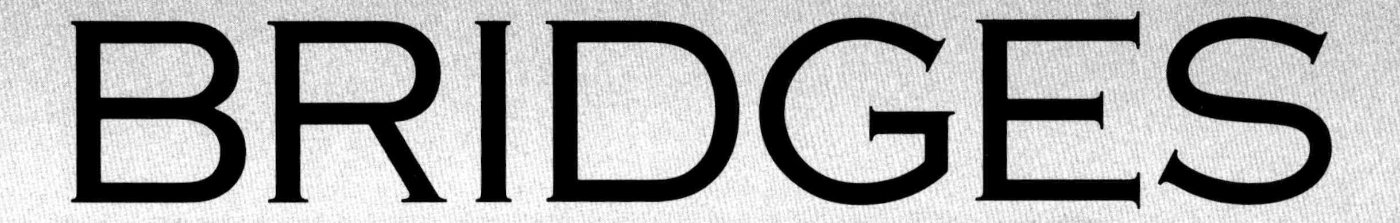

BRIDGES

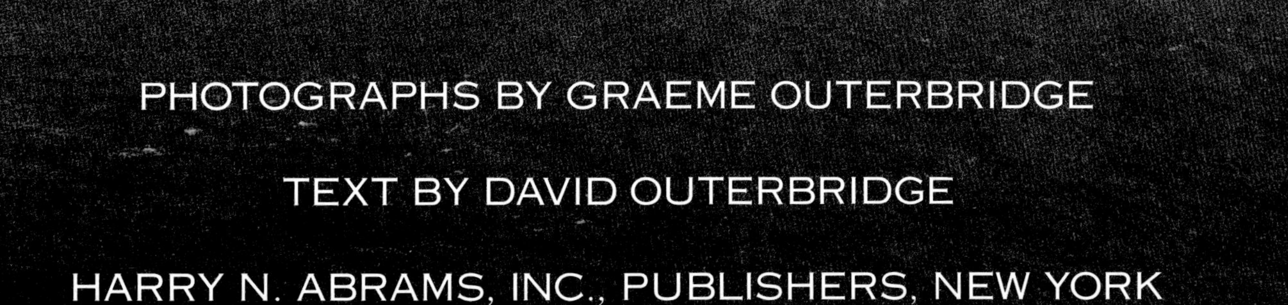

PHOTOGRAPHS BY GRAEME OUTERBRIDGE

TEXT BY DAVID OUTERBRIDGE

HARRY N. ABRAMS, INC., PUBLISHERS, NEW YORK

Editor: Eric Himmel

Design and Montage: Dirk Luykx

Page 1: *Clifton Bridge*
Pages 2–3: *Golden Gate*
Pages 4–5: *Verrazano Narrows Bridge*

Library of Congress Cataloging-in-Publication Data
Outerbridge, David.
Bridges/text by David Outerbridge, photographs by Graeme Outerbridge.
p. cm.
ISBN 0-8109-1239-2
1. Bridges—Pictorial works. I. Outerbridge, Graeme, 1950–
II. Title.
TG149.093 1989
624'.2'0222—dc19 89-318

Published in 1989 by Harry N. Abrams, Incorporated, New York

A Times Mirror Company

Printed and bound in Japan

CONTENTS

ACKNOWLEDGMENTS

I would like to acknowledge with deepest gratitude the following friends, associates, and family, who helped span *Bridges* from idea to reality:

Ricky DeMoura, Dennis Sherwin, Paul Gottlieb, Lilias Outerbridge, Eric Himmel, Dirk Luykx, Terry Falk, Martha Dickinson, Leslie Smolan, Roddy Smith, Rick Smolan, David Cohen, Gerard Bocquenet, Dominique Bocquenet, Andreas Senger, Annetta Sanger, Morris Hayes, Frank Janbourg, Nigel MacIntyre, Gilbert Darrell, Coralita Darrell, Ben Groves, Liz Groves, Stewart Mott, Cappy Wells, Kay Halle, Margaret Sherwin, Susan Swendsen, Greg Swendsen, Ray Gomez, Walter Urbanowicz, Arum Kumar, Margaret Peet, Sam Guilsano, Isabelle Outerbridge, Alexandra Outerbridge, Adam and David Outerbridge, John Kaufmann, Roxy Kaufmann, Flip Schulke, Paula Menefee, Yeaton Outerbridge, Betsy Outerbridge, Alexander Outerbridge, Douglas Outerbridge, Andrew Outerbridge, Alexis Outerbridge, Elizabeth Outerbridge, Louisa Outerbridge, Cecila Mead, Mark Kaufmann, Jean DeGroote, Sabine DeGroote, Caroline Gaty, William Beerman, Robert Mann, Sudie Curtis, Felix Cartenga, Alexander Stuart Outerbridge, Cathy Quealy, Henri, Reggie Cooper.

Graeme Outerbridge

DEDICATION

At the very rim of the Port District there stands, begun during his administration, a physical link binding the two states. The New York piers rest on Staten Island, his first home within the Port District and now his final resting place. Before it was dedicated, the Commissioners of the Port Authority thought it fitting to call it THE OUTERBRIDGE CROSSING. It was intended as a tribute to our first Chairman in recognition of his great service to the Port. It is now a fitting monument to his memory. But greater than the steel structure as a monument is this institution itself. It embodies his spirit, his purpose, and his work.

We here solemnly resolve that to the full extent of our powers, this institution shall carry on and hold its course true to the traditions and standards of EUGENIUS H. OUTERBRIDGE, Chairman April 1921–March 1924.

——The Board of Commissioners, Port Authority of New York and New Jersey

There is a bridge in New York Harbor called the Outerbridge Crossing. It is named for my grandfather Eugenius Outerbridge, to acknowledge his many years of work in creating the Port Authority of New York and New Jersey, which was established in 1921. Completed in 1928, it was the first bridge constructed under the auspice of the new bistate compact. After years of suggesting, negotiating, and establishing the Authority, he served as its first chairman for a salary of $1 a year. When he died in 1932, the *Herald Tribune* wrote: "Men are seldom thus honored during their lifetime by identification with great public works. In the record of New York, Mr. Outerbridge will have an assured rank, typifying the dignity and prestige of the metropolis."

Because of his years of pro bono public service, of which his work creating the Port Authority was the most significant and time consuming, my grandfather died in debt.

More than a half century later Graeme Outerbridge and I would like to dedicate not a bridge but this book about bridges to his memory, to a man who was called by the New York Chamber of Commerce "The Builder of a Great Port."

Only history remembers these things; the world goes on. As the Port Authority's first chairman, my grandfather was given a gold and enamel medallion—the kind of badge police officers flash for identification. It has the Authority crest, his name, and underneath the word "Chairman." A year ago I drove across the Outerbridge Crossing in the course of work for this book. At the tollgate I had the temerity to pull out the medallion to see it if would get me across without a fee. "What the hell is that?" asked the toll collector. I explained the history. "Get your money out. You're holding up traffic," he replied.

David E. Outerbridge

INTRODUCTION

This book is a story of bridges and their evolution. Before I became involved in its research and writing I was quite unaware of bridges as objects. It is impossible to cross, say, the Golden Gate in San Francisco (pages 136–141) without being aware of the graceful span; but that was for me the exception. I took no note of the many bridges I passed over and under almost daily. Bridges were merely an unnoticed convenience that allowed me to get where I wanted to go.

All that has changed. Now I see bridges.

I fly into Newark Airport. On the approach I see dozens of bridges connecting the various islands and high grounds of the New York urban sprawl. I see that the Hackensack River is crossed by several bridges built by competing rail lines. Highway bridges cross the Arthur Kill to Staten Island. Numerous small bridges take roadways over the New Jersey swampland. Driving in from the airport I see the distant beauty of the Bayonne arch (pages 108–111; until recently the largest bridge of its type in the world). I see the long complex of trusses that take rail lines across Newark Bay and the Passaic River. I see bridges of various types that open, built for a day when river traffic was more active. I try to understand the purpose of all the struts, cables, and towers.

By now I know some secrets. I know, for example, that for many of these bridges the difficulty was not the engineering of a span of road or rail bed but building the foundations. The challenge was to dig through hundreds of feet of river mud to solid rock. And once bedrock was struck, to pour concrete and then anchor a pier that could bear the weight of the entire structure. I know that many men were killed or crippled working under rivers in dangerous conditions.

I know other secrets. I know why some bridges are span, some arch, some suspension. (I hardly knew the difference between them before.) I know the tragic stories of important bridges that collapsed, often new bridges. I know what the engineers of those bridges had not yet discovered. I know the origin of bridges.

I want to tell the story of bridges. If I can tell it as I hope I can, the reader will also know these secrets, and bridges will no longer be unseen. They will be noted as—in the words of one historian—an invention as important as the wheel.

MAN AND THE BRIDGE

The oldest engineering work devised by man, it is the only one universally employed by him in his pre-civilized state.
—Joseph Gies, *Bridges and Men*

Imagine a time when there were no bridges. No bridges except those natural outcroppings of rock that formed arches over valleys. (Such crossings still exist: the Pont d'Arc, a 194-foot span across the Ardeche in France, and the rock arch in Rockbridge County, Virginia, which has a span 90 feet long and 140 feet wide, towering 215 feet above Cedar Creek.) Compare that time with today, when nations,

Landwasser Viaduct

states, counties, and cities are connected by patchworks of bridges. *In the city of New York alone there are 2,025 bridges.* Bridges for pedestrians, trains, automobiles, subways, and trucks. Our modern economies and societies cannot exist without bridges.

Man, if not nomadic by nature, is at least migratory, and it cannot have been long in his tenure on the planet before he discovered how to make a bridge. It is a reasonable assumption that bridges predate the wheel and fire because they were already there and could be seen in use by animals. A tree fallen across a stream afforded dry passage, while monkeys using vines to swing between trees provided a different model. The tree was a span-type bridge: a single piece of material that rested on both sides of the stream. The vine was a suspension-type bridge, anchored to, and hanging from, two fixed objects.

In all of mankind's history, technology has produced only one other type: the arch bridge. Span, suspension, and arch are still the three methods we have of bridging a space between two points of ground.

The ways in which bridges are constructed, however, have advanced considerably both in terms of engineering and the development of stronger materials. Today, there are single-span bridges, continuous-span bridges, truss-span bridges, cantilever bridges, a variety of arch bridges, and suspension bridges. There are bridges that combine more than one of these types. There are also bridges that open, which are nothing more than span bridges that can be moved.

Yet in all of this bridge construction, if you stop to think about it, there is no great diversity of invention, only the refinement of technique. Span and suspension bridges existed in prehistoric times. The arch was much used by Roman bridge builders and known to Egyptian and Etruscan builders hundreds of years earlier. Aristotle described diving bells and air hoses two thousand years before James Eads and John Roebling used caissons to build their bridge foundations. Portland cement was invented in 1823, but concrete was well known to the Romans, who used a local substance, pozzolana, mixed with sand that would harden under water into a firm substance. Several ancient cultures used iron in bridges centuries before Thomas Telford in England built his first bridge of iron in the eighteenth century.

About the only thing that has changed is that modern materials permit longer bridges to be made with lighter materials. And we have done away with the slave labor that built many notable ancient structures. We still kill people in the building of bridges, however; our bridges also still kill people. The musselmen of the River Tay in Scotland who hooked bodies out of the eddies below the bridge of that name after its collapse in December 1879 have their modern counterparts.

BEAUTY OF STRUCTURE

This thing, beauty, requires a few words.

In the objects that man makes for use, form is inherently related to function. When the maker of an object has found the most appropriate form to meet its intended use, beauty will follow. This criterion of beauty—the unity of form and function—can be applied to most of the things man makes, whether they be pitchers, saddles, boats, or bridges. The difference between a pitcher and a bridge, however, is the difference between craft and science. The craftsman works with an intuitive knowledge of his materials. (The great bridge builder Eugene Freyssinet also relied on the artisan's intuition. He is, however, an exception.) An engineer applies calculation to the known physical properties of his materials.

A perfect pitcher will be of a clay or with a glaze that makes it watertight. The spout will pour cleanly without an after-drip. It will stand firmly, and when it is full the handle will provide the proper leverage for pouring. Its wall will not be thicker than necessary, and it will contain a more generous quantity of liquid than expected. It will be the essence of utility, hence, it will embody beauty. It may well have been made from a lump of clay in less than two minutes.

To build a bridge, an engineer must understand physics: the dynamics of compression, tension, and torque. He must know the tensile strength, elasticity, and brittleness of his materials. He must be able to do the mathematical calculations that measure weight and stress. The engineer with these tools can design a structure that will serve its purpose. An element of beauty will come from the bridge's conformity to its environment, but it can only be realized if the form is true to the function. The goal is similar to that of the craftsman but is reached through calculation elevated by an intuitive aesthetic.

There was a time when the distinction between craftsman and engineer was less clear. The man who built a canoe or wherry was working within both disciplines. And early bridge-builders, farmers, for example, who needed to move cattle and grain across a stream, were also somewhere between the world of craft and engineering. They needed to get from here to there and used their knowledge and ingenuity to fashion a structure that would accomplish the crossing.

Utility, function, and efficiency do not mean that beauty must be stark or even cold. Grace and elegance are part of the equation. Embellishment may also be incorporated in the design, but this is dangerous ground.

Our lives are littered with objects made hideous by inappropriate ornamentation. By and large, however, our great bridge builders have been chary of the application of elements not essential to structure. Any loss of aesthetics we note in bridges is usually due to an overuse of material, such

Ponte Vecchio

as the potter's overthick walls of a pitcher or a gaudy and unnecessary design on the glaze. For some earlier bridges, their designers may be forgiven an overly cumbersome structure: Not knowing precisely the dynamics of their materials, they chose to overbuild.

The Forth Bridge in Scotland (pages 30–33) is an example of such overbuilding. Following the collapse of the Tay railroad bridge, to err on the side of safety must have been in the engineers' calculations. In fact they allowed for a wind pressure of 50 pounds per square inch against the structure when 30 pounds is known today to be sufficient. Nonetheless, the excess is only in the dimensions of the bridge's parts; there is a beauty in the bridge's overall mighty presence, a union of elements. Everyone must be his own judge of that aesthetic, however. Quite a different verdict was rendered by William Morris. Morris, a nineteenth-century Romantic and disciple of Gothic purity, found the bridge hideous. "There never would be an architecture in iron, every improvement in machinery being uglier and uglier until they reached the supremest specimen of all ugliness, the Firth of Forth Bridge."

THE CRITERIA FOR A BRIDGE

Why is New York's Verrazano Bridge of the suspension type? Why the Forth Bridge a cantilever? Why is any bridge precisely the design it is? The answers have to do with these six basic elements:

—Length of span required
—Load the bridge must bear
—Terrain to be bridged
—Longevity of bridge
—Available economic resources
—Available materials and state of technology at the time of construction

Length of span. What is the length required of the span? Obviously, the distance to be bridged determines the parameters of the bridge. Before the Industrial Revolution this question was less critical. With the exception of viaducts that had to move water continuously, there was no compelling need other than convenience to span great distances. Most bridges cross water, and before development of the railroad with its capacity to move heavy freight, boats could move what traffic existed quite satisfactorily. Boats could move armies, and did, for that matter. Only the Romans, being consummate bridge builders with an ample labor force, spanned rivers for conquest. The Romans were ahead of their time as bridge builders, and after the collapse of their empire no bridge of significance was constructed in the West for eight hundred years.

Length of span has been the primary limiting factor in bridge building. It is the reason, for example, that the Hudson River between New York and New Jersey was not spanned until 1931, and when it was, the George Washington Bridge that crossed it was 50 percent longer than any suspension bridge ever built. The 3,500 feet of water it crosses could only have been spanned by a suspension bridge. Continuous span or arch bridges could only have been constructed with a frequency of foundations that would have ended river navigation, and at a cost beyond calculation.

Load. Bridge engineers divide the weight of a bridge into two quantities: live and dead load. The dead load is the actual weight of materials used to construct the bridge. Added to dead load is the weight of traffic (live load) that the bridge is being designed to carry. This latter factor changed radically with the development of the railroad. Locomotives and freight cars are substantially heavier than horse-drawn vehicles and pedestrian traffic. Consider that a single pair of train wheels on an axle weighs 2,000 pounds. Consider that the weight of many cars, perhaps even the entire train if the bridge is long, have to be supported. (The effect of wind force is also a calculation of load.)

Trains present an additional problem: The load is dynamic. As an example of how this affects bridge stability, imagine the following: It is one thing to carry a bride across the threshold. Presumably she is peacefully accepting the transportation. It would be another thing to try and carry a kicking, thrashing person a similar distance, or at all. As anyone who has traveled by rail knows, a train has considerable swaying motion. It also tends to rise and fall on its springs as it speeds over the rail bed. Both of these motions add stress, which is equivalent to weight, to the bridge structure. On the whole, the problem presented by dynamic loads was understood quite quickly by builders because they knew railroads. From the invention of the steam locomotive until the 1860s, almost all railroad bridges were designed by railroad engineers, as opposed to engineers trained in the structural design of bridges themselves. This dynamic load phenomenon was also known from the action of animals and people on bridge crossings. Soldiers marching in cadence have collapsed bridges, as have herds of swaying cattle. "The pounding of horses' hoofs . . . were enemies above all to be dreaded."

The cantilever and trussed-span-type bridges were well suited to the load-bearing requirements of railroad bridges, which is why so many bridges of the nineteenth century were built of that type. With the advent of the automobile and truck, ever so much lighter than trains, the cantilever was by and large replaced as a bridge type in favor of suspended roadways.

Terrain. Terrain also affects the engineer's choice of design. A short deep ravine might be bridged by a single span or arch, anchored at each end. This simplifies the construction, as foundations and piers do not need to be built.

Trains introduced an additional problem that had not

Ganter Bridge

been a factor in the day of animal power or more recently with automobiles. Trains can move enormous weight at high speeds, but they cannot travel over any but the mildest of inclines. Trains do not go uphill. Thus, an uneven terrain had to be bridged to establish a near-level rail bed. Miles and miles of bridges, many of them intricate trussed spans, cross our central states to enable grain, coal, and cattle to move over the land.

One aspect of the effect of terrain on bridge design concerns the navigation requirements of the water to be spanned. Waterways vital to river traffic cannot be bottled up by a mass of foundation work. This raised problems for engineers in earlier times, who had neither the knowledge nor materials to construct long spans under which ships could pass. For this reason some needed bridges were not built; others, like the old London Bridge, changed economic and social patterns; still others, overextended, collapsed.

Longevity. We tend to think that bridges are built to last forever. Such is not always the case, most notably in the case of military bridges. Many early bridge builders were not engineers at all, but travelers who needed to get across a body of water that was not fordable. A simple span that might be replaced the next season if need be was adequate to their needs. Suspension bridges in Asia, made with vines, were not expected to last; if one bridge gave way, another was built to replace it.

Economics. The availability of both labor and money has always affected bridge design. A vast labor supply was the only reason that the Romans were able to translate their exquisite understanding of the arch into massive structures, such as the Pont du Gard in Nîmes, France (pages 64–65), and the Puente Alcántara that bridges the River Tagus in Spain, that stand to this day.

Often bridge construction has been financed by bonds that are repaid by subsequent tolls charged on the bridge. (In earlier times, bridges also used to charge tolls for traffic that passed under them.) Unfortunately, once a bridge was paid for, tolls were often discontinued. Many current bridge closings are due to the fact that money was never levied for their maintenance.

Materials. Although length, load, and longevity are all factors that become intertwined in decisions concerning bridge construction, the availability of materials has dominated consideration of how to build a bridge through all ages.

Until the industrial age, wood and masonry were the primary building materials. Then, in a very short time, a half dozen new materials became available that radically enlarged man's ability to bridge distances. In rough order of appearance these materials were: cast iron, wrought iron, concrete, steel, reinforced concrete, alloy and silicon steel, and prestressed concrete. Of these, only the last is not common in our equipment and possessions and merits a brief explanation here.

Concrete, like stone, is strong in compression and weak in tension. That is to say, it is able to resist pressure against it but has little ability to withstand pressures exerted to pull it apart. Reinforced concrete, developed in the late nineteenth century, is a material in which the ability of metal to withstand tension was added to concrete's ability to withstand compression. Iron or steel rods or beams were encased in concrete to combine the two strengths. Anyone walking by a construction site will have observed this combination—rods sticking out of concrete. It is, however, a static strength, whereas prestressed concrete possesses a dynamic one, the metal within acting like a spring. Prestressed concrete is made by casting a concrete beam with longitudinal holes. Steel cable is threaded into these spaces and then tightened into a stretched mode. The ends of the taut cables are then anchored into the ends of the concrete. This creates a force that is constantly trying to pull the beam together. This contained energy can be used to counterbalance the opposing forces in an arch that are pressing down and out. The strength of this material has allowed engineers to create bridges using a small volume of material, which in turn creates an appearance of considerable grace. (The Salginatobel Bridge in Switzerland on pages 106–107 is a good example.)

FOUNDATIONS

Except for spans of short distances that can be bridged by a single girder resting on both banks, all bridges that span rivers and bays require foundations whose footings are under water. These foundations have been the bane of bridge designers and their laborers for all time, and their construction over the centuries has inspired great ingenuity and cost dearly in human lives. Typically, one-half of the expense of a bridge is consumed in foundation construction.

The stone "clapper" bridges of England with their monolithic spans were admirable achievements of "primitive" people who knew how to move stones of enormous weight. However, their foundations in the streams they crossed were pretty simple constructions; witness the bridge in Dartmoor, England (pages 18–19). What of more substantial flows such as the Euphrates, the Tiber, the Danube, or the East and Hudson rivers of New York?

Aristotle records the use of diving bells and air hoses to allow workers to operate dry under water in building bridge foundations in ancient Greece. But the Greeks were a seafaring people; they had small use for bridges and thus were not compelled to work on their development. The Romans, on the contrary, made their great expeditions on land and had to master the spanning of rivers that were obstacles in their path.

Richmond Bridge

Centuries before either the Greek or Roman civilization had blossomed, about 2,000 B.C., Semiramis, the Queen of Babylonia, bridged the Euphrates. Her engineers were inventive in a most fundamental way. They waited until the dry season then built a dike that diverted the low volume flow of water into a temporary lake. Then they assembled a labor force on the dry riverbed and built stone piers that they bonded with iron bars, which were in turn soldered with lead. The roadway was made of timber, a portion of which could be taken up at night to keep invaders from crossing the bridge. Once completed, the river was diked back to its normal course.

One of the more primitive but effective methods of laying foundations under water was to fill old boats with stones and sink them in place. This was a great improvement over the even simpler system of just throwing rocks into the water and creating a pile on which a span could rest because the boats, until they rotted, helped contain the rocks and prevented currents from rearranging and ultimately destroying the foundation.

The Romans contributed inventions to the construction of bridge foundations that have hardly been improved in two thousand years. Materials and technology have radically changed over the centuries, but the essential engineering solutions of building cofferdams and driving piles remain the same.

A cofferdam is a watertight compartment that is built in the river where a foundation is to be constructed. A series of poles are driven side by side into the riverbed in a square or rectangular shape, with the ends of the poles extending above the water level. The water inside the cofferdam is then pumped out and the space made as watertight as possible. Laborers are then sent into the cofferdam to dig out the mud of the river bottom until a firm bed is reached. A foundation can then be erected within the cofferdam.

When Othmar Amman, the great builder of New York bridges, drew up plans for his towering edifice, the Verrazano Bridge (pages 142–145), he employed precisely the same cofferdam technology. And he accomplished this in the swirling currents of the New York Narrows, working to a depth of 170 feet below the water level. Needless to say, his materials—sheet steel and concrete—were more securely watertight than poles, branches, and clay, but the engineering was the same.

The Romans were skilled at driving piles deep into the river bottom. Roman engineers perfected mechanical pile drivers with capstans and levers. They also discovered that wood that had been charred rotted more slowly, adding to the foundation's life. A concentration of driven piles became a firm foundation.

So refined was this technology that in 55 B.C. Caesar was able to construct a bridge across the Rhine, a distance of 1,500 feet, in just ten days. This feat becomes the more extraordinary when compared to the fact that the length of the Brooklyn Bridge (pages 128–129), which took fourteen years to complete, is only 95 feet greater.

Caesar's legions cut trees 18 inches in diameter. These were sharpened at one end and driven into the riverbed. A pair of these piles was set angling against the stream. Another pair of piles was then driven 40 feet upstream and angled toward the lower set. On and on across the river twin sets of piers were thus placed in position. A wide roadway was then built on the foundations. Caesar dismantled the bridge upon his return, for security, but the remains existed into this century.

In the second century A.D., the Emperor Trajan, also in a mood to conquer, decided to cross the Danube to attack the barbarians. Rather than using boats, he had one Apollodorus of Damascus build a bridge of timber arches on twenty masonry piers that spanned 3,000 feet!

The piers were 60 feet wide, 50 feet thick, and 150 feet high. Consider that. Across one of the world's greatest rivers Roman bridge builders constructed a series of stone towers each the size of a modern ten-story apartment building. As the Roman Empire faltered to a close, a later emperor had to destroy the bridge to keep the barbarians out.

What is truly noteworthy about Trajan's bridge across the Danube is that it was not for 1,200 years that a greater span was constructed anywhere in the world, and it was 1,600 years before the Danube was bridged again.

When we build a house we rightly think of the foundation as the base on which everything else's stability will depend. Foundations of bridges serve the same purpose, but they are not, as are building foundations, in a tranquil environment. Typically, they are in a moving flow of water. Water itself will wear away a foundation over time. This process of erosion is known as scour. (Consider, after all, that it is only water of the Colorado River that has carved the Grand Canyon.)

The water itself may carry ice and debris that smash against a foundation. Aside from scour and erosion, currents exert forces on entire structures. Every year thousands of piers are destroyed along our shorelines as the sea carries away the pilings. Bridges, too, are vulnerable.

To minimize these dangers, bridge engineers protect foundations with cutwaters, also a Roman invention. A cutwater is simply an angled construction built around a foundation that shears the oncoming stream, forcing it by with a minimum impact on the structure. French bridge builders, in particular, created very graceful cutwaters that added to the overall beauty of their bridges.

Nothing, of course, can protect a foundation from the impact of a vessel driven against it, and bridges have collapsed with loss of life when struck at the foundation by ships passing underneath.

Just how difficult the building of a foundation can be was demonstrated in the construction of the railroad bridge

Iron Bridge

over New York's Hell Gate. There, in the ever-moving maelstrom of water that churns through the Narrows, engineers discovered a chasm 60 feet wide running along the river bed. Unfortunately, a pier had to be placed at this location, and test drillings could find no bottom to the mud with which it was filled. In order to create a solid base for the foundation Gustav Lindenthal, the engineer, had to design a gigantic concrete arch spanning the crack and capable of withstanding violent currents, and workmen had to construct it 70 feet below the surface of the East River. The foundation for the Hell Gate Bridge was then built on this arch, carrying on up above the river to bear what was then the heaviest bridge in the world. This foundation was designed to sustain a load of 52,000 pounds of dead weight *per linear foot* of rail bed and another 24,000 pounds per foot of moving trains. The Hell Gate Bridge weighed more than Lindenthal's two other East River bridges combined, the Manhattan and Queensboro bridges (the latter being until then the heaviest bridge in the world).

In the 1870s, a new way to dig through mud on the river bottom was developed: the pneumatic caisson. A caisson (from the French word *caisse,* meaning box) is a huge wood, metal, or concrete box without a bottom. The caisson is lowered to the river bottom; its top remains above water. Air pressure inside the caisson keeps water out. A work force of laborers, nicknamed "sandhogs," enters the caisson through an air lock and with pick and shovel dig away at the soft river bottom. As the hole deepens, the caisson sinks lower. Additional weight in the form of concrete is added to drive it downward, and the air pressure inside is increased to equal the increased water pressure of the lower depth.

Magere Bridge

At the time James Eads was constructing the bridge that bears his name in St. Louis and John Roebling had finished his design for the Brooklyn Bridge, no one knew the medical consequences of being placed in an environment of compressed air. Today we know that someone who has been under forced pressure must only gradually return to normal atmospheric pressure. A rapid decrease in pressure results in "the bends," a painful and often fatal affliction caused by nitrogen releasing itself in the body in the form of bubbles.

Because of ignorance of this, dozens of men were killed in the construction of the foundations of these two bridges. The malady was called the "caisson disease." Sandhogs were considered to be highly paid for this wretched work—$2 a day on the Brooklyn Bridge. However, the work was so miserable that approximately a third of the caisson workers quit each week.

There was also the danger of blowouts, which happened when compressed air found a way to escape under an edge of the caisson. One blowout during construction of the Brooklyn Bridge created a waterspout almost 200 yards high. Others blew men up into the air to their deaths.

Once the excavation process has lowered the caisson to rock, it is filled with concrete and becomes the base for a column.

The pneumatic caisson is rarely used anymore. The working conditions are too dangerous, and exposure to the pressure necessary for some foundations would mean a man could only work two thirty-minute shifts in a day yet be paid a full day's wages.

Instead, an open caisson can be used in which both the bottom and portions of the top are open. Cranes with buckets dig out the river bottom through openings, and the caisson, built with sharp bottom edges, sinks down as the excavation proceeds and weight is added to the top. The Verrazano Bridge was constructed in this manner.

Another recent development in foundation construction has been to employ the principle of buoyancy. Huge concrete air chambers filled with water are lowered into place and anchored. The water is then pumped out and replaced with air. This could be viewed as an adaptation of the principle of the pontoon bridge, where the roadway is supported by floating metal pontoons. (As these buoyant concrete chambers support up to 30 percent of a bridge's weight, the puncture of them will likely make good material in an espionage thriller some day.)

DEATH ON THE BRIDGE

London Bridge is falling down,
Falling down, falling down.
London Bridge is falling down,
My fair lady.

—Nursery Rhyme

In 1887, George Vose, studying bridge disasters in this country, wrote: "Not less than forty bridges fall in the United States every year."

Bridge disasters continue to happen.

Within the past decade in the United States bridges have washed away, bridges have fallen down, bridges have been deemed unsafe and closed. Lives have been lost in these accidents. More lives have been lost in the building of bridges. And more lives will be lost. A death for every $10,000,000 of bridge expense has been one calculation of the expected fatalities in modern-day bridge construction.

The most notorious bridge disaster of the nineteenth century occurred on Sunday, December 28, 1879, at 7:20 P.M., when the Edinburgh Mail, carrying seventy-five passengers, was midway across the Tay Bridge at Dundee, Scotland. Suddenly both it and the bridge were gone.

The collapse had been predicted by one Patrick Matthew, locally known as the Seer of Gourdie: "In the case of accident with a heavy passenger train on the bridge the whole of the passengers will be killed. The eels will come out to gloat over in delight the horrible wreck and banquet."

It was also an accident that could have been predicted by

engineers, had they known what was revealed in the subsequent inquiry.

Amazingly, Thomas Bouch, the bridge's builder, had not taken into account wind pressure against the bridge itself and the additional effect of wind against a passing train. The Tay was well known for violent storms, and the tragedy occurred during one of them. Ordinarily, wind force is a primary calculation for its effect on a bridge and is considered a part of its load, especially where the bridge has a bulky structure, or sail area.

The builder also had not accounted for the percussion effect of the transit of heavy train wheels clacking over the tracks and the vertical movement created by the passage of a train. Shortly before the accident, a maintenance worker had noticed approximately 300 pounds of bolts that had popped from their girders due to this percussion.

Lastly, the bridge had been constructed with substandard materials. Defective holes in the girders had been skillfully masked with "Beaumont's Egg," a blend of beeswax, iron filings, rosin, and lampblack. It seems unimaginable that anyone involved in the fabrication of a bridge would knowingly supply defective structural materials. Yet the Tay Bridge is not an isolated example. Portions of the wire cable that holds the Brooklyn Bridge up were discovered to be faulty after the cables were in place. However, Washington Roebling, its chief engineer, anticipating substandard materials somewhere in the construction, had compensated for that in the design by overbuilding.

Clifton Bridge

The most dramatic bridge collapse in this century did not claim any lives but did reveal the effect of oscillation upon a suspension bridge. As the roadway of this type of bridge literally hangs from the cables that support it, certain forces can cause it to start swinging. A rank of soldiers marching in cadence can start an oscillation. This movement, which may be exacerbated by wind, can build enough motion so that the bridge tears itself to pieces. In a report on the Tacoma Narrows Bridge collapse, engineers noted: "At the higher wind velocities torsional oscillations, when once induced, had the tendency to increase their amplitudes."

The Tacoma Narrows Bridge, south of Seattle, nicknamed "Galloping Gertie" during its short life, provided a vivid display of oscillation. The roadway was prone to start an undulating, wavelike motion. Fascinated with the amusement-parklike thrill, motorists drove for miles to experience the excitement of the wild drive across its span during which the road might undulate up and down by as much as 5 feet.

On November 7, 1940, a strong wind started some alarming undulations in which the left and right sides of the road were rolling in a contrary twisting motion. As the rate of oscillation slowed, the amount of distortion increased. Soon the motion was so violent that one edge of the road would first be 28 feet above the other, then 28 feet below! A photographer and an engineer who had been on the bridge crawled desperately back to shore along the centerline. Then the bridge tore itself apart and plunged into the water below. (This brought an end to a local bank's promotional billboard, which read "Safe as the Tacoma Bridge.")

Bridge designers learned something from this. Othmar Amman, who had built the George Washington Bridge a decade earlier without stiffening trusses, wrote: "The Tacoma Narrows Bridge failure has given us invaluable information." The George Washington Bridge, because of its great weight, 56,000 tons, did not require stiffening, but lighter bridges like the Tacoma span were subsequently designed with girders or stays that prevented oscillation, and existing bridges were modified.

If Amman's comment on the disaster seems a bit callous, it can be compared to that of Isambard Brunel, who reputedly responded to the news that one of his bridges had collapsed: "I am very glad. I was just going to build a dozen like it."

The most common cause of bridge collapse is faulty design. Perhaps it was inevitable with the great mercantile and industrial need to span ever greater distances that human error would take its toll. The great Quebec cantilever bridge collapsed before completion in 1907, sending 9,000 tons of steel and seventy-four workers to the bottom of the St. Lawrence River because the design was too fragile in the first place and warning rents in trusses already in place were ignored.

Future bridge failures will likely be less due to bad design than to poor maintenance, but there will be failures, and there will be continued death on the bridge.

NODAL AND SOCIAL BRIDGES

The old London Bridge was replaced in the nineteenth century. But by then it had done its work for more than six hundred years as a nodal bridge, the catalyst that created the city of London. Work on the bridge was begun in 1176 by one Peter of Colechurch. Because of the difficulty of constructing piers in the flowing Thames it was not completed until 1209. (More than two hundred people were killed during construction, most by drowning.)

Once the decision had been made to build the bridge, however, the evolution of what would be London was forever changed. Because the bridge effectively closed the upper river to seagoing shipping, it drastically changed traffic patterns. Thus, it is what social scientists call a nodal bridge. Had the London Bridge not created a node on the navigable Thames, London would not have become the city it is. When finally completed, the bridge had sufficiently blocked the river that water surged through the arches like a millrace, with a great roar. The bridge created a point of terminus for cargo coming up the river from overseas.

Roads, warehouses, and later a rail system, all centered in London, and merchants of every variety settled in London to direct the distribution of goods all across England and, later, down the Thames and across the Empire.

The London Bridge was also a social bridge. Its wide roadway was bordered with buildings. Multi-storied houses and shops were joined at rooftop and formed tunnels.

The London Bridge became a high-rent district in part because of the expeditious sewage disposal. People shopped and lived there, and pedestrians enjoyed watching the water coursing through the arches. The bridge was also a place for earthy punishments meted out by the Crown: Thieves hanging on hooks and the heads of criminals on display were part of the bridge's adornments.

One historian described a memorable spectacle: "The Pope having sent John Fisher, Bishop of Rochester, a cardinal's hat, Henry [King of England] commented, 'Mother of God, he shall wear it on his shoulders for I will leave him never a head to put it on.' Cardinal Fisher's head was parboiled before being displayed above the tower gate on the bridge. But it refused to decay, and after two weeks during which the martyr's face grew ever more shining and attracted ever larger crowds of awed believers, the executioner was ordered to throw it into the river by night. Sir Thomas More's head replaced it. . . ."

The London Bridge was not the only bridge of an earlier time that provided opportunity for shopping and leisure. The Rialto in Venice, with its arcades of small shops, and the Ponte Vecchio in Florence, with concessions specializing in jewelry, are other examples included in this book.

Although the latter still has commerce conducted over its arches, these bridges have lost their social purpose. The marketplace as a center of social intercourse is largely of the past, and "market day" is extinct in our culture. We move too fast. Pedestrian traffic had a different meter than automobile life. On these European bridges there was once time to tarry, conclude a bargain, and then relax. Taverns, even bordellos, used to exist on these bridges.

One of the best descriptions of the social time on a bridge comes from Arthur Upham Pope, who wrote about the Pul-i-Khaju that crosses the Zaindeh at Isfahan. It was typical of a Persian seventeenth-century travel stop, a more leisurely version of today's interstate accommodation: "Poets in contemplation, mullahs and philosophers in argument, families in reunion, gossiping women and friends at various amusements, all may enjoy the beauties of the scene in isolation and contentment."

BRIDGE BUILDERS

We know the names of almost none of the Roman bridge builders, which is a shame, for no culture before or after Rome has ever produced such a body of work in bridge construction.

We do know, however, that the Romans did have a group of priests who specialized in bridge engineering. They were known as the Collegium Pontificium (college of bridge makers), and their head was called Pontifex Maximus. This tradition has carried on at least in nomenclature into Christian times, and the Pope is today still referred to in that way, Pontifex Maximus.

After the decline of the Roman Empire, bridge building appears to have continued to be a clerical pursuit. In the Middle Ages a group of Benedictine monks (called the Pontist Friars by British historians) is credited with expertise in bridge engineering, and although there is some historical disagreement, it seems likely that the Saint Bénèzet who built the Pont d'Avignon was of that order.

Some earlier bridge builders merit brief mention for their signal accomplishments.

The famous Rialto bridge of Venice (pages 74–77) was built by Antonio da Ponte in 1588. (His last name translates into English as "of the bridge.") Da Ponte faced much political opposition over his commission, and people doubted that the bridge would survive on the soupy landscape on which it stood. Da Ponte, however, drove 6,000 birch and alder trees as piles into the mud at each abutment, and the Rialto stands on them to this day. (Da Ponte was also an avid volunteer fireman who once saved the Doge's Palace when he discovered it ablaze.)

Jean-Rodolphe Perronet (1708–1794), often called the "father of modern bridge building," transformed the design of the arch in bridges. David Steinman, himself a noted bridge builder of this century, has called Perronet "one of the most illustrious engineers the world has ever known." Curiously, he is rarely included in encyclopedias of biography or general knowledge.

Perronet's career was helped by a childhood encounter. In a playground in Paris's Tuilleries one day he was invited by the young Prince Louis XV to take part in a game. A friendship was born that led to Perronet's appointment as director of the Royal Corps de Ponts et Chaussées. In the course of a long life of bridge construction, he fundamentally changed the understanding of arch dynamics.

Perronet's finest bridge, the five-arch span at Neuilly-sur-Seine, exists no more. However, the Pont de la Concorde in Paris does. Unfortunately, he was not permitted to follow completely his own design and had to modify it in ways he found unaesthetic and unnecessary. Already an old man at the time of the bridge's inception, he died in his sleep in a shack at the construction site.

Thomas Telford (1757–1834) was born in Scotland. Home was a one-room thatched cottage. Telford had little formal schooling and spent part of his youth as a shepherd. He was a cheerful lad, known as "Laughing Tam." Befriended by a wealthy spinster, he taught himself everything he could from the books in her library. He then learned

Salginatobel Bridge

masonry as a hewer, apprentice, and journeyman. Excited by the radical American Thomas Paine's 400-foot, cast-iron span (cast in England for Pennsylvania's Schuylkill River, but never erected) Telford started designing bridges with the new material.

Telford bridges the traditions of craftsman and engineer; and also the ages of masonry and metal in bridge use. His iron bridge across the River Spey (pages 90–91) is the first modern metal arch using principles associated with iron as opposed to imitating the architecture associated with stone. His iron suspension bridge at Menai (pages 122–125), while employing stone arches on the land approaches, was the longest suspension bridge in the world when built, and was a model for suspension bridges for a century.

Poor Telford built more than one hundred and twenty bridges but was never able to enjoy retirement. Late in life he wrote, "Having by now been occupied for about seventy-five years in incessant exertion, I have for some past time arranged to decline the contest; but the numerous works in which I am engaged have hitherto prevented my succeeding." And even in death he may not have attained his wish. He left instructions to be buried in a quiet chapel, but the Society of Civil Engineers, which he founded, would have no such thing. Instead he is buried next to the trample of feet in Westminster Abbey.

George Stephenson (1781–1849) and his son Robert (1803–1859) are important in the evolution of the railroad and its bridges. The father was born of a family that was dirt poor, son of a fireman on a steam pump. He, like children the world over, dreamed of being an engineman. Instead he became a cobbler to support himself, and it was not until he was eighteen that he learned to read and write. He then became adept at repairing engines, was for a time a brakeman, and then "engine wright." He invented the "Rocket steam locomotive" that could recycle its steam for additional power and speed.

George Stephenson tried to get his son the education he had missed, and, in 1850, Robert built the first railroad bridge of boxed girder form in which the train actually passed through a rectangular hollow. In 1852, he was honored by Queen Victoria and Prince Albert, who walked through the bridge. George Stephenson, like Telford, is buried in Westminster Abbey.

Leon Moissieff (1872–1943), unfortunately, is best remembered for the collapse of his bridge at the Tacoma Narrows.

Gustav Lindenthal (1850–1935) was born in Austria and emigrated to the United States, where he trained as an engineer. In part he learned his trade as what was called a "computer" for the Keystone Bridge Company.

Lindenthal's imprint on bridge design is best seen in New York. He designed the cantilever Queensboro span (pages 40–43), the Manhattan suspension bridge further down the East River, and the rail crossing over the Hell Gate. David Billington, who has written extensively about bridge construction, said of the latter, "His Hell Gate Bridge, coming at the end of railroading's dominance, symbolized the era with all its pretension, ambiguity, and power."

Joseph Strauss will be remembered for the extraordinary beauty of the Golden Gate Bridge in San Francisco (pages 136–141). He should also be noted for the great lengths he went to in order to ensure the safety of his workers. He even commissioned special eyeglasses to counter the sun's reflection off San Francisco Bay. And he ordered sauerkraut juice for workers who arrived for work with hangovers.

The Hudson River is a mighty body of moving water, ice-choked in winter, that has scoured a channel deep in the rock palisades of its New York and New Jersey shores. It and the New York Narrows, up which most of the immigrants to this country traveled, were the last major unbridged waterways in the New York harbor. Was it possible to span them?

Othmar Amman (1879–1966) bridged both. He also built the Goethel Narrows Bridge, the Outerbridge Crossing (pages 50–53), and the Bayonne Bridge (pages 108–111), linking New York City and New Jersey, the Triborough Bridge across the East River, and the Bronx-Whitestone Bridge from The Bronx to Long Island.

The Swiss-born Amman was a major engineering consultant to Lindenthal on the Hell Gate Bridge and for decades was the chief engineer of the Port Authority of New York. That he should have been able to build the George Washington Bridge (pages 130–135) virtually simultaneously with several other structures is unprecedented. That he built three bridges to dimensions never before known is extraordinary. That he is himself not immortalized by a great bridge named in his honor, a shame.

Eugene Freyssinet (1879–1962) was the pioneer of prestressed concrete and its deployment in light, graceful spans. Freyssinet was born of humble farmer parents, and throughout his professional life he was inspired by the traditions of country craftsmen, the economies of smiths and joiners. "For him only two sources of information existed: direct observation and intuition, in which he had more faith than in calculation," writes one biographer. Freyssinet himself observed, "We never find more at the end of a calculation than we put in it at the beginning." And about the craftsmen whose ideals he cherished and deployed in bridge building, he wrote, "These men have created for themselves a civilization the main characteristic of which is an extreme concern for the simplification of forms and economy of means."

In the pages that follow are bridges built by many of these men. There are also bridges whose builders are mentioned only on the page on which they appear. The feeling of some of these bridges, such as the Pulteney Bridge (pages 80–81), is quiet. Others, such as the Ganter Bridge (pages 58–61), possess a soaring beauty. All, however, have been selected because their builders created a form that well served function.

Menai Straits Bridge

SPAN

The simplest of the three bridge types—span, arch, and suspension—is the span.

A span is a beam that is supported at each end by a foundation, which may mean nothing more than that it lies on the ground. The force exerted by a load on a span is one of gravity alone, a downward pressure on the foundation:

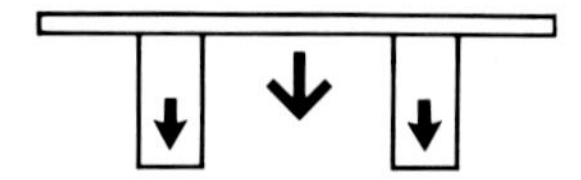

Spans have been made from stone, wood, iron, steel, concrete in its various forms, and from various combinations of these materials.

Although simple in concept, the span bridge can be the most complex in appearance because a great many supporting members—struts and trusses—may be added to make the span sufficiently strong and stable. For an example of the appearance of complexity, look at the Forth Bridge (pages 30–33).

A span does not have nearly the strength of an arch or a suspension bridge. This is because a weight placed upon it creates a force of compression on the top and tension on the bottom:

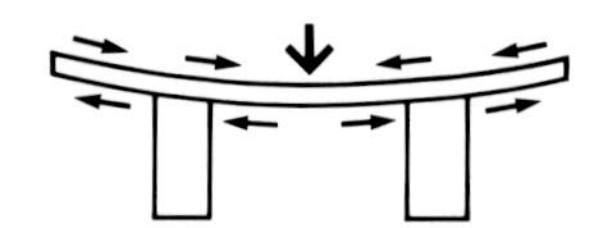

Thus, stone, which is strong in resisting compression, will crack underneath. And metal, which is strong in its ability to resist tension, will react to the compression force and bend.

Consequently, if the distance to be bridged is great, a series of intermediary foundations will be required. A bridge constructed of more than one span resting on shared foundations is known as a continuous-span bridge.

Then there are truss-span bridges. Anyone who has built almost anything will understand the purpose of a truss. If you nail together a rectangle of two-by-fours and then put pressure on any of its corners, it will be distorted into a parallelogram. However, if you nail a diagonal truss between the corners, it will be impossible to distort the shape because the triangles created by the diagonal truss are stable:

(This was the principle of the Bailey bridges of World War II: Spans capable of carrying tanks and other heavy equipment could be quickly erected out of many small metal triangles.)

The cantilever bridge is a special type of span construction, a span that is anchored at only one end. "Walking the plank," the pirate's supposed form of execution, used the cantilever principle. A piece of wood was extended out over the side of a ship, the inboard end being lashed down to the deck. The condemned wretch was sent out on the plank and off its end to his doom.

Another way of visualizing the cantilever is to think of a seesaw. When in balance, the board rests horizontally on its fulcrum. If one end of the seesaw is then weighted or anchored to the ground, the free end is able to carry weight without any underlying

BRIDGES

support. Engineers often construct cantilever bridges in this way, building out simultaneously from a central point until the inner side reaches a shore and can then be anchored to the ground.

Not all cantilever bridges are "pure" cantilever. Bridge designers discovered that if they built a cantilevered span out from either shore there was sufficient strength in the structure to connect the two sides in the middle with a suspended span between them.

The final connection can be a matter of some difficulty, requiring very precise planning. Because metal expands and contracts with temperature, it is sometimes a tortuous process to get the necessary alignment, in which holes for fastening come into position to be bolted. This problem of making connections was well illustrated in the construction of the Forth Bridge's final connection.

The Forth Bridge (pages 30–33) is the earliest long span built on the cantilever principle. It was an extraordinary achievement for its day. The Firth of Forth is a raw body of water, throwing up huge seas, with an almost 20-foot tide range. The caissons used to dig the foundations were 90 feet high and 70 feet in diameter. Construction took seven years, cost forty-four lives, and required 140,000 cubic yards of masonry and 44,000 tons of steel.

Finally there remained only the last connection to be made, tying the two ends of the Forth Bridge together in the middle. However, as the weather remained cold, the metal had contracted and a gap of some inches remained. The engineers waited, and then on an October day the sun shone clear on the bridge and the gap started to close. The temperature was still 5 degrees below what was needed, and a frustrating ¼-inch space remained to be closed. The engineers ordered the steel beams to be swathed in cotton, which was soaked in oil and set afire. Then, ever so slowly, the great cantilevers moved toward each other until there was a fit, and the fastenings were set in place.

The opposite problem was faced by James Eads in his construction of the arched-span bridge that crosses the Mississippi River at St. Louis. An unusually warm spell of weather expanded the top portions of an arch beyond each other by more than 2 inches. Workers had to build wooden casings around the steel. These were packed with more than 100,000 pounds of ice. This cooled the metal, which then contracted to a point where the arch could be closed.

Bridges that open are made of spans that move.

A drawbridge raised a span over a protective moat. A bascule bridge operates on a similar principle, with a counter-weight attached to the end of the span that goes down. Thus, when the end of the span over the channel is raised to let a ship pass beneath, it requires only a little energy. There are bascule bridges where the balance is so accurate that the bridge keeper can open the bridge by hand.

There are also vertical lift bridges where the entire span is raised in a horizontal position, also made practical by the use of counter-weights. And there are swing bridges, where the span rotates on its foundation, opening like a gate.

There are rolling bridges where the span is pulled back on wheels. There are a few transporter bridges (pages 44–47), which are actually more like tramways than bridges. Traffic is moved on a platform, suspended from cables, that rolls along a permanent span fixed high above a river.

Finally, there have been "sinking" bridges. These rare spans are built so that they can be lowered to allow river traffic to navigate past them overhead.

NAME *Post Bridge* LOCATION *River Dart, Dartmoor, England* TYPE *Span*
LENGTH *Each slab app. 15′* MATERIAL *Stone* DESIGNER/ENGINEER *Unknown*
DATE COMPLETED *c. 1,000 B.C.–A.D. 1*
REMARKS *Stone "clapper" bridges such as this one dot the English countryside. Their dates range from pre-Roman through medieval times.*

NAME *Magere Bridge* LOCATION *Amstel River, Amsterdam, Holland* TYPE *Drawbridge*

LENGTH *Span 33′; overall length 276′* MATERIAL *Wood*

DESIGNER/ENGINEER *Unknown* DATE COMPLETED *c. 1670; widened 1840, 1934*

NAME *Portage Viaduct* LOCATION *Genesee River, Portageville, New York* TYPE *Span*
LENGTH *818′* MATERIAL *Iron*
DESIGNER/ENGINEER *George S. Morison* DATE COMPLETED *1875*

NAME *Smithfield Bridge* LOCATION *Monongahela River, Pittsburgh, Pennsylvania* TYPE *Span*
LENGTH *Two spans 360′ each* MATERIAL *Steel*
DESIGNER/ENGINEER *Gustav Lindenthal* DATE COMPLETED *c. 1883*

NAME *Forth Bridge* LOCATION *Firth of Forth, Scotland* TYPE *Cantilever span*
LENGTH *Main spans 1,710′; overall length 5,330′* MATERIAL *Steel*
DESIGNER/ENGINEER *John Fowler and Benjamin Baker* DATE COMPLETED *1889*
REMARKS *Visually, this is one of the most powerful bridges ever built.*

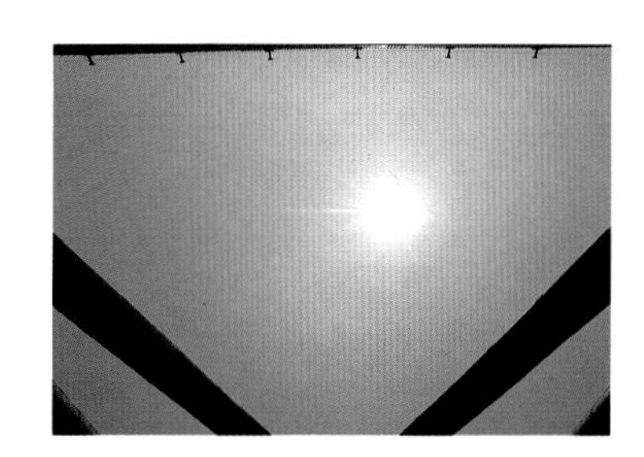

NAME *Walnut Street Bridge* LOCATION *Susquehanna River, Harrisburg, Pennsylvania* TYPE *Span*
LENGTH *3,604′* MATERIAL *Iron*
DESIGNER/ENGINEER *Phoenix Bridge Company* DATE COMPLETED *1890*

NAME *Queensboro (or 59th Street) Bridge* LOCATION *East River, New York, New York*
TYPE *Cantilever span* LENGTH *Span 1,182′; overall length 7,449′* MATERIAL *Steel*
DESIGNER/ENGINEER *Gustav Lindenthal* DATE COMPLETED *1909*

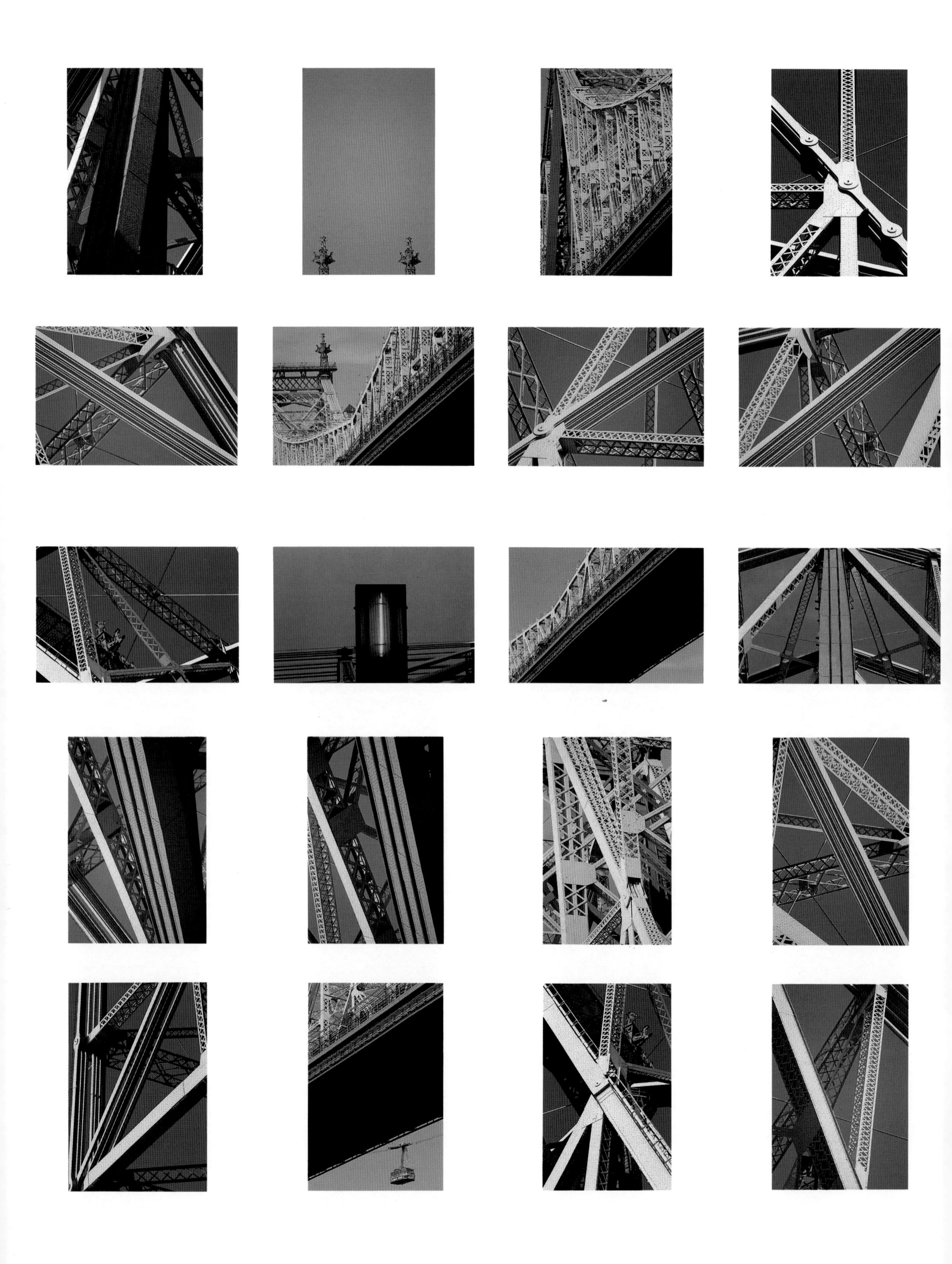

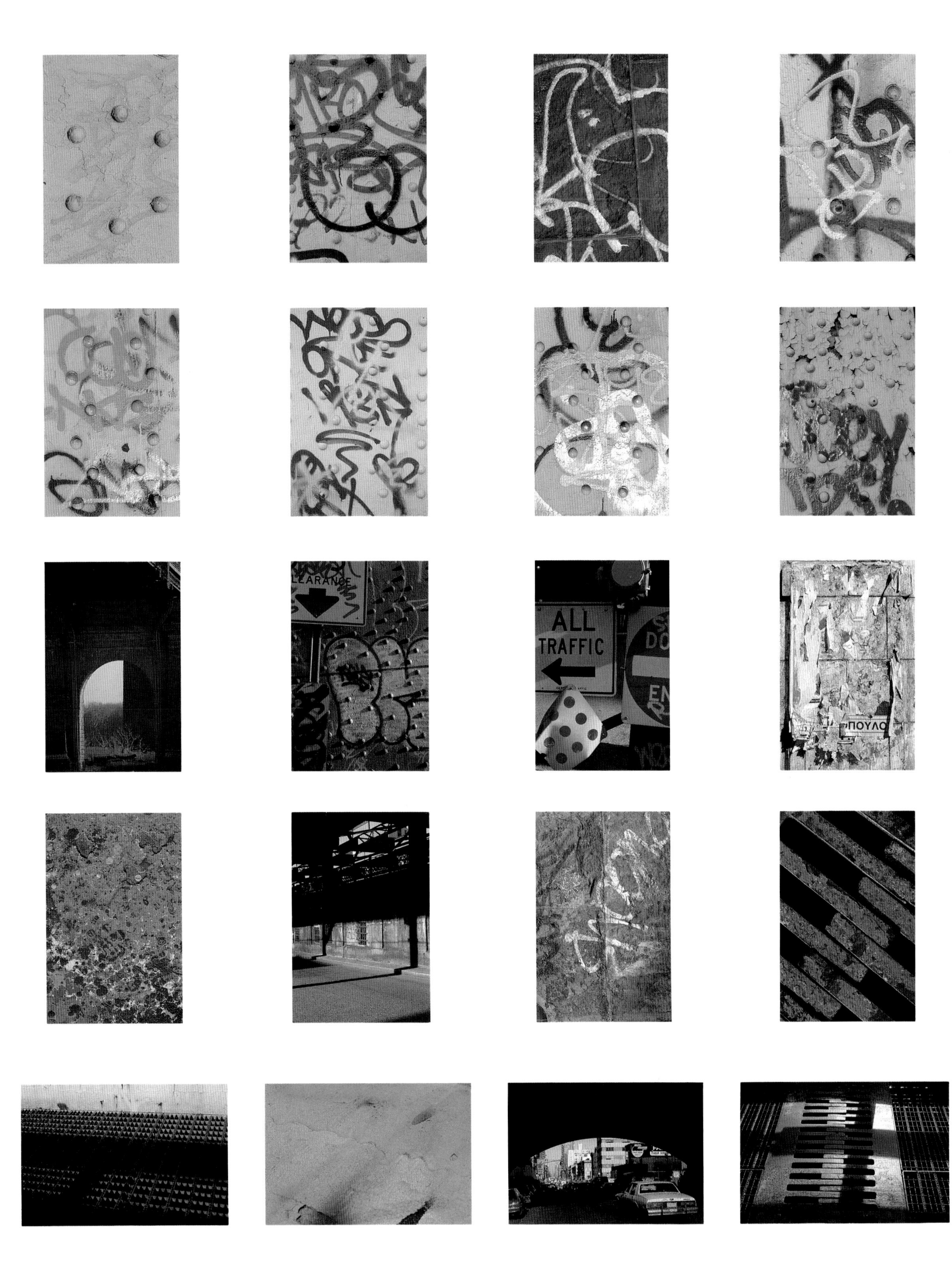
ALL
TRAFFIC
ΠΟΥΛΟ

NAME *Pittsburgh & Lake Erie Bridge* LOCATION *Ohio River, Beaver Falls, Pennsylvania*
TYPE *Span* LENGTH *Main span 769′; overall length 1,787′* MATERIAL *Steel*
DESIGNER/ENGINEER *Albert Lucius* DATE COMPLETED *1910*

NAME *Middlesbrough Transporter Bridge* LOCATION *River Tees, Middlesbrough, England*
TYPE *Span* LENGTH *Span 470′* MATERIAL *Steel*
DESIGNER/ENGINEER *Cleveland Bridge & Engineering/Sir William Arrol & Co./S. Arnodin/William Pease*
DATE COMPLETED *1911*

STOP

NAME *Sciotoville Bridge* LOCATION *Ohio River, Sciotoville, Ohio* TYPE *Span*
LENGTH *1,550′* MATERIAL *Steel*
DESIGNER/ENGINEER *Gustav Lindenthal* DATE COMPLETED *1917*

NAME *Outerbridge Crossing* LOCATION *Arthur Kill, between Perth Amboy, New Jersey, and Staten Island, New York* TYPE *Cantilever* LENGTH *Main span 750′; overall length 10,140′* MATERIAL *Steel* DESIGNER/ENGINEER *Othmar Ammann* DATE COMPLETED *1928*

NAME *Burlington Northern Railway Bridge* LOCATION *Hanover, Montana*
TYPE *Span* LENGTH *1,392'* MATERIAL *Wood*
DESIGNER/ENGINEER *H. S. Loeffler* DATE COMPLETED *1930*
REMARKS *This bridge is typical of the many trussed spans built to accommodate the railroad's need for a level roadbed.*

NAME *Ganter Bridge* LOCATION *Ganter Valley, Switzerland*

TYPE *Cantilever span* LENGTH *Main span 477′* MATERIAL *Prestressed concrete*

DESIGNER/ENGINEER *Christian Menn* DATE COMPLETED *1980*

REMARKS *New materials and form unite into one of the world's most graceful and beautiful structures.*

ARCH

The development of the arch in architecture and bridge-making allowed for a fundamental improvement in the load-bearing capacity of structures, not to speak of aesthetics.

The arch is a bridge type of its own because it distributes load forces differently than a span. Unlike a span, the arch is not normally used as a carriage way but as the support for one. Depending on the construction, an arch bridge may have the roadway run over it, through it, or beneath it:

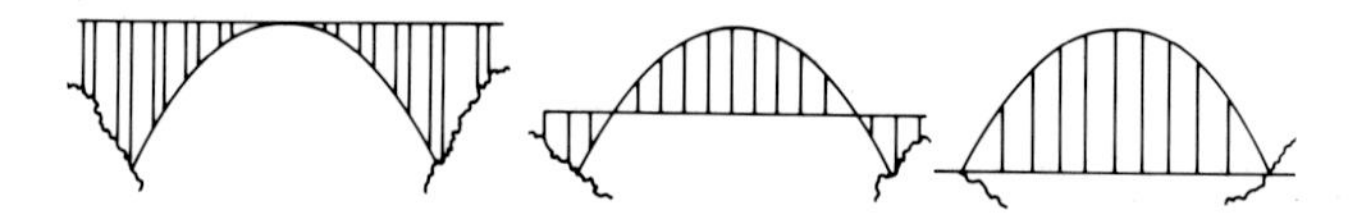

What makes the arch different in principle from a set of upright piers perpendicular to the ground are the forces that are exerted by weight upon it. In the case of a simple span, the force is solely gravitational. In the case of the arch, this load is spread through the principle of compression, where the downward force is directed outward to the abutments of the arch, thus spreading the pressure and increasing the carrying load:

The Roman arch was a semicircle. Because of this shape it could normally span a distance four or, at most, five times the width of its supporting columns. Following the Roman Empire it took several hundred years before an improvement of the arch evolved in the West. The Chinese had the camelback arch bridge—short spans with an exaggerated rise to allow river traffic to pass underneath. (On these bridges the arch *was* the roadway.) Today, there are many arch types: semicircular, elliptical, bowstring, and basket-handle, to name a few.

The great aqueduct at Nîmes (pages 64–65) was built by the order of Agrippa, son-in-law of Augustus Caesar, in 19 B.C. It has a length of 885 feet and runs 155 feet above the valley floor. In order to construct the arches (as in almost all masonry work involving arches) falsework first had to be erected. Falsework is the web of wooden timbers that holds the stones in place during construction until the final wedge-shaped center stone is installed. The arch then supports itself and the falsework is removed. Looking at the Nîmes aqueduct with its three tiers of arches, one marvels at the vast amount of timbers that would have to have been cut and assembled to accomplish the task.

Equally extraordinary is the fact that the Romans cut and pieced their stones so carefully that no mortar was required to hold them together. Another Roman bridge, the Puente Alcántara over the Tagus River in western Spain, whose 98-foot arches rise 170 feet over the river, was constructed in A.D. 105 of stones that weigh eight tons apiece. It seems incredible that its builder, Caius Julius Lacer, could have accomplished this in the middle of a turbulent river. But he got the job done, and of his bridge he said, "Pontem perpetui mansuram in saecula mundi" (I have built a bridge that will last for eternity).

So far he has been proven correct. During the Peninsular War, the French did once attempt to destroy it in battle but only succeeded in taking out one arch. Wellington, in pursuit, used ship's rigging and timber to erect a suspension span across the gap that was able to accommodate his forces. It should also be noted that when repair work was done on

BRIDGES

the bridge in more modern times, the engineers were unable to do their job without the use of mortar.

In the twelfth century an inspired friar in Avignon decided to build a bridge across the Rhône (pages 66–67). Only a portion of it remains today, but its builder (Saint Bénèzet) was the first to design a three-centered arch. Thus, the Roman semicircle was somewhat flattened and broadened.

In the eighteenth century, French engineer Jean-Rodolphe Perronet made a fundamental discovery in arch dynamics. He calculated that the horizontal thrust, that is, the force of compression outward from the top of the arch, would carry through, and be countervailed by, the force of an abutting arch. As the live load would be distributed through all the arches to the final abutments on the banks, it meant that each pier only needed to support the dead load of the arches resting on it. As a consequence, the ratio of arch to pier could be changed. Thus he was able to design a bridge where the arch was broadened and the pier narrowed to achieve a ratio of 9:1 arch span to pier compared to the Roman ratio of 4:1.

Engineers and architects talk of the more pleasing aesthetics of these gracefully proportioned gentle arches. Perronet's bridge over the Seine at Neuilly-sur-Seine (no longer in existence) was considered one of the most beautiful in the world. His discovery also had more useful functions: With fewer foundations more widely spaced, river navigation was easier, there was less damming pressure on the river, and less erosion of the foundations.

The first iron-arch bridge (pages 86–89) was built by Abraham Darby III in 1779 at Coalbrookdale in England, the home of a famous iron foundry. It weighed a mere 378 tons and stands to this day. Over time the earth embankments have moved closer to each other, forcing the arch upward and creating a slight point at its peak. Darby's Iron Bridge was followed by Thomas Telford's 1814 cast-iron bridge over the River Spey in Scotland (pages 90–91), which, unlike its predecessor, took advantage of the new material to reform arch design out of its masonry tradition, that is to say the new material did not substitute for stone; it allowed a design appropriate to iron, which was lighter in mass visually and in fact.

In 1874, James Eads built the first great arch bridge of steel in St. Louis. A railroad bridge, it was to be a symbol of the gateway to the West. In his letters and petitions on its behalf to the city, Eads always referred to it as "your bridge." Each of the three arches has a span of more than 500 feet, which was unprecedented. Eads, who knew the Mississippi River from years of salvage work on it, knew less about bridge building. Yet he was a most scrupulous inspector of its materials. One of his suppliers, the young Andrew Carnegie, complained of the rigorous testing and commented that Eads had rejected so much steel that the bridge could have been built of silver at the same cost.

In France, at almost the same time, Gustave Eiffel, whose name is forever tied to his Paris tower, was building the Garabit Viaduct (pages 92–95), a single-arch bridge, not of steel but wrought iron. When completed, the graceful span of 453 feet was the largest of its type in the world.

The development of reinforced and prestressed concrete has allowed arches of even shallower curves to be built. In our time, from the drawing boards of such architects as Eugene Freyssinet and Christian Menn (pages 114–115), the arch has been transformed into a perfection of visual and functional unity.

A visual sweep. The spanning of space. Beauty.

NAME *Pont du Gard* LOCATION *Nîmes, France* TYPE *Arch* LENGTH *885′*
MATERIAL *Stone* DESIGNER/ENGINEER *Unknown* DATE COMPLETED *A.D. 19*
REMARKS *Dominating its landscape, this Roman bridge was built of stones so carefully hewn that no mortar was required.*

NAME *Pont d'Avignon* LOCATION *Rhône River, Avignon, France* TYPE *Arch*
LENGTH *Partly destroyed; spans of four remaining arches 101–110' each* MATERIAL *Stone*
DESIGNER/ENGINEER *Saint Bénèzet* DATE COMPLETED *Twelfth century*
REMARKS *Only a portion of the bridge remains standing.*

NAME *Ponte Vecchio* LOCATION *Arno River, Florence, Italy* TYPE *Arch*
LENGTH *Central span 100′; side spans 90′; overall length 390′* MATERIAL *Stone*
DESIGNER/ENGINEER *Taddeo Gaddi* DATE COMPLETED *1345*

NAME *Rialto* LOCATION *Venice, Italy* TYPE *Arch* LENGTH *Span 89′* MATERIAL *Stone*
DESIGNER/ENGINEER *Antonio da Ponte* DATE COMPLETED *1588*
REMARKS *Each abutment rests on 6,000 birch and alder pilings driven deep into the soft ground.*

VENEZIA E LA DIFESA
DEL LEVANTE
G. CA

NAME *The Palladian Bridge* LOCATION *Stowe House, Buckinghamshire, England*
TYPE *Arch* MATERIAL *Stone* DESIGNER/ENGINEER *William Kent* DATE COMPLETED *c. 1734*

NAME *Pulteney Bridge* LOCATION *Avon River, Bath, England* TYPE *Arch* MATERIAL *Stone*
DESIGNER/ENGINEER *Robert Adam* DATE COMPLETED *1770*

NAME *Dolan-Hirion Bridge* LOCATION *Towy River, Llandovery, Wales*
TYPE *Arch* LENGTH *Span 84′* MATERIAL *Stone*
DESIGNER/ENGINEER *William Edwards* DATE COMPLETED *1773*

THOMAS EDWARD
1773

NAME *Richmond Bridge* LOCATION *Thames River, Richmond, Surrey, England*
TYPE *Arch* LENGTH *Spans 45′, 50′, and 60′; overall length 299′* MATERIAL *Stone*
DESIGNER/ENGINEER *James Payne* DATE COMPLETED *1777*

NAME *Iron Bridge* LOCATION *Severn River, Coalbrookdale, England* TYPE *Arch*
LENGTH *100′* MATERIAL *Cast iron*
DESIGNER/ENGINEER *Abraham Darby III* DATE COMPLETED *1779*
REMARKS *This was the first iron bridge. More than fifty years would elapse before an iron bridge was built in the United States.*

AND ERECTED IN THE YEAR MDCCLXXIX

ERECTED
IN
1779

NAME *Craigellachie Bridge* LOCATION *River Spey, Elgin, Scotland*
TYPE *Arch* LENGTH *152′* MATERIAL *Cast iron*
DESIGNER/ENGINEER *Thomas Telford* DATE COMPLETED *1814*
REMARKS *This was the first iron bridge engineered to use the qualities of iron in design, as opposed to iron mimicking masonry and carpentry principles.*

NAME *Garabit Viaduct* LOCATION *Trugère River near Garabit, France*
TYPE *Arch* LENGTH *453'* MATERIAL *Wrought iron*
DESIGNER/ENGINEER *Gustave Eiffel* DATE COMPLETED *1884*
REMARKS *At the time of its completion, this was the longest arch span in the world.*

NAME *Glenfinnan Viaduct* LOCATION *Inverness-shire, Scotland*
TYPE *Arch* LENGTH *1248′* MATERIAL *Stone*
DESIGNER/ENGINEER *Simpson & Wilson/Robert McAlpine* DATE COMPLETED *1898*

NAME *Pont Alexandre III* LOCATION *River Seine, Paris, France*
TYPE *Arch* LENGTH *353′* MATERIAL *Steel*
DESIGNER/ENGINEER *Louis-Jean Résal* DATE COMPLETED *1900*

NAME *Landwasser Viaduct* LOCATION *Albula Pass, near Filisur, Switzerland*
TYPE *Arch* LENGTH *446′* MATERIAL *Stone* DATE COMPLETED *1904*

NAME *Tunkhannock Viaduct* LOCATION *Tunkhannock Creek, Nicholson, Pennsylvania*
TYPE *Arch* LENGTH *2,375′* MATERIAL *Concrete*
DESIGNER/ENGINEER *Abraham Cohen* DATE COMPLETED *1915*

NAME *Market Street Bridge* LOCATION *Susquehanna River, Harrisburg, Pennsylvania*
TYPE *Arch* LENGTH *3,618′* MATERIAL *Concrete*
DESIGNER/ENGINEER *Modjeski & Masters* DATE COMPLETED *1928*

NAME *Salginatobel Bridge* LOCATION *Schrau River, Schiers, Switzerland*
TYPE *Arch* LENGTH *Span 246′* MATERIAL *Reinforced concrete*
DESIGNER/ENGINEER *Robert Maillart* DATE COMPLETED *1930*

NAME *Bayonne Bridge* LOCATION *Kill van Kull, between Bayonne, New Jersey, and Staten Island, New York* TYPE *Arch* LENGTH *1,652′* MATERIAL *Steel*
DESIGNER/ENGINEER *Othmar Ammann* DATE COMPLETED *1931*
REMARKS *Amazingly, Ammann completed this bridge, the longest arch in the world at the time, in the same year he completed the George Washington Bridge, the longest suspension span.*

NAME *Rainbow Bridge* LOCATION *Niagara Falls, New York*
TYPE *Arch* LENGTH *950′* MATERIAL *Steel*
DESIGNER/ENGINEER *Aymar Embury II* DATE COMPLETED *1941*

NAME *Reichenau Bridge* LOCATION *Rhine River, Reichenau, Switzerland* TYPE *Arch*
LENGTH *Span 291′* MATERIAL *Prestressed concrete*
DESIGNER/ENGINEER *Christian Menn* DATE COMPLETED *1964*

NAME *Fehmarnsund Bridge* LOCATION *Fehmarnsund, West Germany* TYPE *Arch*
LENGTH *Span 813′* MATERIAL *Steel and concrete*
DESIGNER/ENGINEER *T. Jahnke/P. Stein/G. Lohmer* DATE COMPLETED *1963*

SUSPENSION

Among modern bridges, the suspension type is the most readily identified, and has become almost a stereotype of "bridge." Typically, it consists of a parallel set of cables that are strung between towers and anchored at each shore. From these cables hangs a network of wires that supports the roadway that lies between the cables. The Golden Gate (pages 136–141), Verrazano (pages 142–145), George Washington (pages 130–135), and Brooklyn (pages 128–129) bridges are classic examples. However, to contemplate more primitive suspension structures is to see the engineering principle more easily.

The simplest form of a suspension bridge was a single piece of rope attached at sufficient height on each bank of a stream or river. The rope was made of braided vines, bamboo, animal hide, or other flexible fiber. One end of the bridge was secured to a tree or rock while the other was floated or carried across to the opposite shore. It was then pulled taut and secured. This bridge exists in a state of tension, which is the principle of all suspension structures.

For this primitive suspension bridge, the method of crossing was also primitive. A person wrapped his legs around the rope and, hanging underneath, pulled himself hand over hand across. From time to time as the fibers stretched, the rope needed to be tightened.

A modest improvement in passenger comfort resulted when a sheave was attached to the rope. This, greased with yak butter or some appropriate lubricant, made it possible to roll along underneath for part of the way. Sometimes there would be two bridges next to each other that, by the way they were suspended, would allow a downhill passage for each direction.

A breakthrough in suspension bridge construction came with the multiple-rope bridge. Two ropes were strung for holding onto, and a third strand was placed underneath for the feet. This improvement was followed by the placement of two or more ropes underfoot with either netting or boards between them to facilitate walking. Many bridges of this type still exist in Asia. Westerners who have journeyed across them have described it as a terror-filled adventure. The bridge sways in the wind and the flooring cants with each step. The rope handrails buck and twist with the traveler's movement. (It is these very motions, caused by wind and the movement of the live load, that modern engineers have had to reckon with in their great structures.)

Because of the tension placed on the vegetable or animal fibers, and the decaying influence of sun and rain and frost, these bridges did not have long lives. However, as both the labor and material needed for them was readily available and inexpensive, the cables could be replaced at frequent intervals. Some bridges snapped while in use. As a defense against attack, the cables could be hacked through. Kipling's wonderful story, "The Man Who Would Be King" climaxes as Daniel Dravot, proud roguish spirit to the end, walks out on the suspension bridge that leads to the Kingdom of Kafiristan. "Out he goes, looking neither right nor left, and when he is plumb in the middle of those dizzy dancing ropes, 'Cut, you beggars,' he shouts; and they cut, and old Dan fell, turning round and round and round, twenty thousand miles, for he took half an hour to fall till he struck the water, and I could see his body caught on a rock with the gold crown close beside."

To go from these simple suspension bridges to their modern equivalents required a change not in principle, but a change in materials. James Finlay, a justice of the peace in

BRIDGES

Fayette County, Virginia, patented a design for a suspended level-floor bridge supported by metal cables in 1808. His first bridge of this suspension design had a span of 70 feet. He built it for $600, and threw in a fifty-year warranty. In the next decade he built forty more bridges of this type. The metal suspension bridge had arrived.

Although other iron suspension bridges had preceded it, the Menai Straits Bridge in Wales (pages 122–125) is often considered to be the first modern metal structure of this type. At the time of its completion in 1826 it became the longest span—1,710 feet—in the world. It was built by Thomas Telford, who used iron bars linked together to form the cables, which were carried over massive stone towers. These towers and the shallow catenary of the cables influenced bridge design for a century.

Everything about Telford's work was done with meticulous care. The wrought iron was tested to a fracture point of almost 90 tons for a 35-ton load. Each link of the cables was cleaned, heated, bathed in linseed oil before assembly, and then painted with oil-based paint. Lewis Carroll, having fun with events of the day, wrote in *Alice in Wonderland,* "I heard him then, for I had just/Completed my design/To keep the Menai Bridge from rust/By boiling it in wine."

Because it was not until iron, and later steel, became available that suspension spans could be made long, the evolution of this bridge type was far slower than the increasingly long span-type bridge. Today, the suspension type has become the most commonly employed for bridging long distances; even as a continent away its rope counterparts continue to serve the transportation needs of isolated populations.

Because of their generally graceful—even fragile—profiles, modern suspension bridges conceal their strength.

The suspension cables of the George Washington Bridge (pages 130–135) are built of 26,474 parallel steel wires. The four cables of the Verrazano Narrows Bridge (pages 142–145) each contain 26,106 wires. Laid end to end, the wire of just these two New York bridges would be long enough to circle the equator more than ten times. These wires are as thick as pencils and when spun together form a round cable with a diameter of about 3 feet.

While primitive suspension cables could be floated or swam across a stream, modern bridges require far more sophisticated equipment. The cables for the Menai Straits Bridge were hoisted from a barge by a capstain turned by one hundred and fifty laborers. The first Niagara Suspension Bridge at Niagara Falls (no longer in existence) was started with a kite contest. A prize was given to the first person who could fly a kite across the river to the opposite bank. The kite string was then used to pull a rope across, and it a hawser, and it a wire cable. Today, most cables are "spun" by machines that cross back and forth from shore to shore carrying new strands of wire.

All of the weight of the bridge and its traffic is carried by the cables over the towers to the two shores. Here is where the strength of a suspension bridge is hidden, literally buried. Tension is constantly trying to pull these cables from their fastenings in the ground. To prevent this, the designed anchorages are truly massive. Today, they are made of concrete blocks as much as 100 feet square, set deep in the ground. Inside the blocks are heavy steel beams that in turn anchor the large eyes to which the tens of thousands of cable wires are individually secured.

NAME *Menai Straits Bridge* LOCATION *Menai Straits, Wales*
TYPE *Suspension* LENGTH *Span 570′* MATERIAL *Wrought iron*
DESIGNER/ENGINEER *Thomas Telford* DATE COMPLETED *1826*
REMARKS *Telford's suspension design influenced subsequent bridges for a century.*

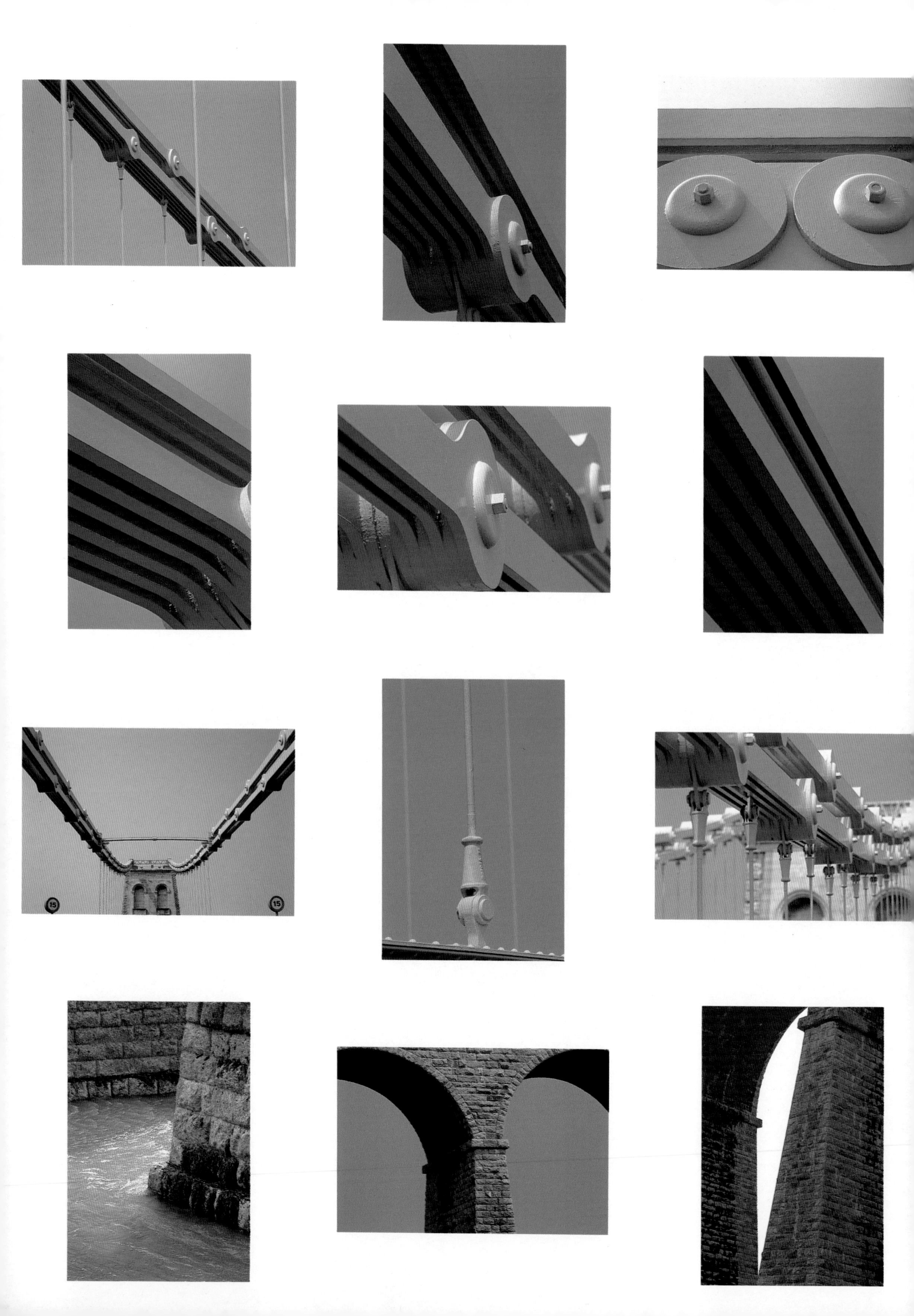
15
15

NAME *Clifton Bridge* LOCATION *Avon River, Bristol, England*
TYPE *Suspension* LENGTH *702′* MATERIAL *Iron*
DESIGNER/ENGINEER *Isambard Kingdom Brunel* DATE COMPLETED *1864*

NAME *Brooklyn Bridge* LOCATION *East River, New York, New York*
TYPE *Suspension* LENGTH *Span 1,595'; overall length 5,989'* MATERIAL *Steel*
DESIGNER/ENGINEER *John Roebling* DATE COMPLETED *1883*
REMARKS *At the time of its completion, the Brooklyn Bridge became the world's longest span.*

NAME *George Washington Bridge*
LOCATION *Hudson River, between New York, New York and Fort Lee, New Jersey*
TYPE *Suspension* LENGTH *3,500′*
MATERIAL *Steel*
DESIGNER/ENGINEER *Othmar Ammann*
DATE COMPLETED *1931*
REMARKS *The bridge is unique in design, containing no storm stays or stiffening trusses. Its great weight provides stability. Le Corbusier called it "the most beautiful bridge in the world. . . . It is blessed. It is the only seat of grace in the disorderly city."*

NAME *Golden Gate* LOCATION *Golden Gate, San Francisco, California*
TYPE *Suspension* LENGTH *Main span 4,200′* MATERIAL *Steel*
DESIGNER/ENGINEER *Joseph B. Strauss* DATE COMPLETED *1937*
REMARKS *The Golden Gate followed the George Washington Bridge in record-setting length. It was not until 1965 that its span was exceeded (by 60′) by the Verrazano Narrows Bridge.*

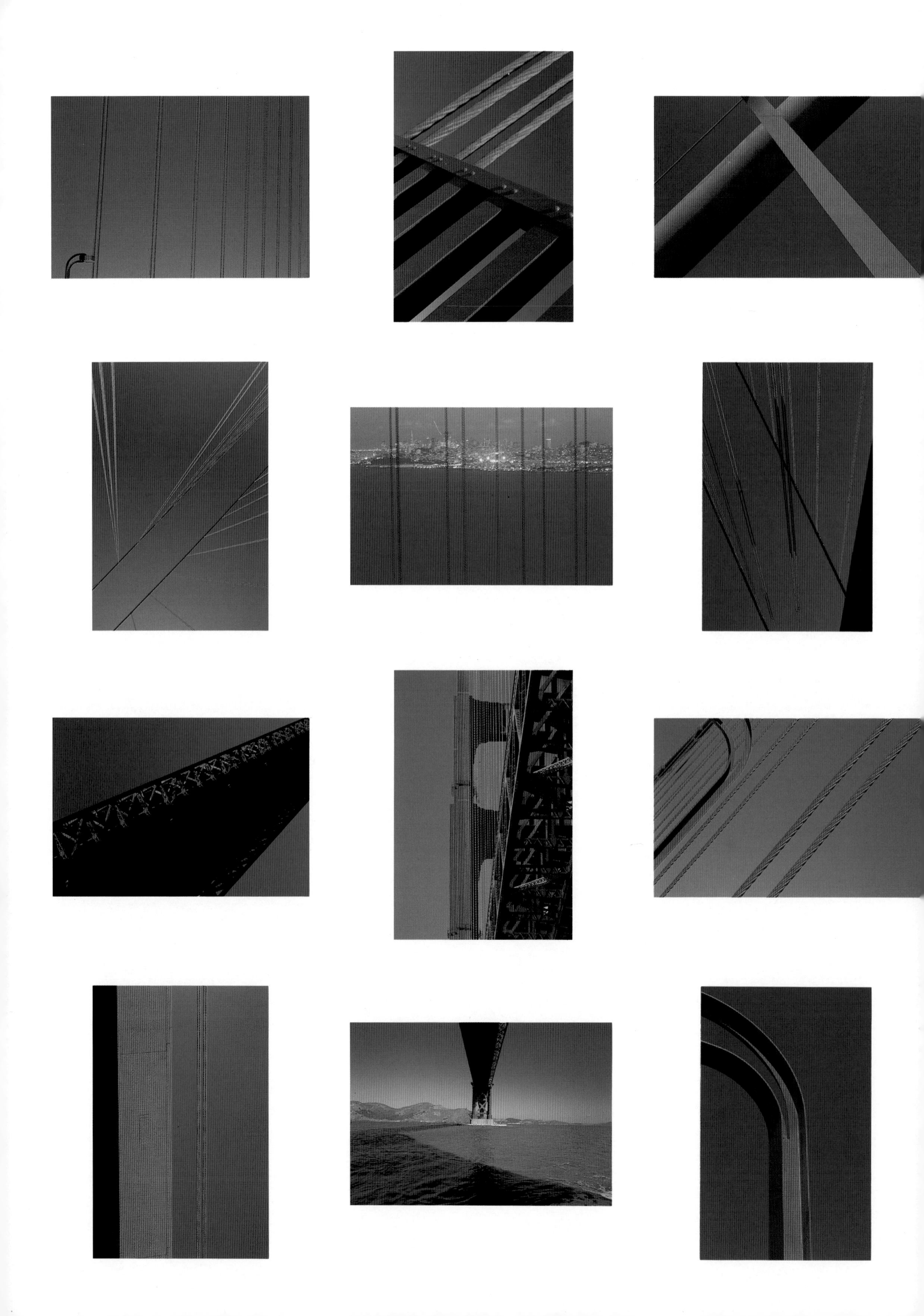

NAME *Verrazano Narrows Bridge*

LOCATION *Verrazano Narrows, New York, New York*

TYPE *Suspension* LENGTH *4,260′*

MATERIAL *Steel*

DESIGNER/ENGINEER *Othmar Ammann*

DATE COMPLETED *1965*

REMARKS *This is the longest suspension span in the world, and the great bridge builder Ammann's final achievement.*

NAME *Rhine Bridge* LOCATION *Rhine River, Flehe, West Germany*

TYPE *Suspension* LENGTH *Main span 1,207'; overall length 1,797'*

MATERIAL *Steel and concrete*

DESIGNER/ENGINEER *H. Grassl/G. Dittmann/R. Kahmann/G. Lohmer/Dyckerhoff & Widmann*

DATE COMPLETED *1979*

NAME *Rhine Bridge* LOCATION *Rhine River, Speyer, West Germany* TYPE *Suspension* LENGTH *Main span 902′; overall length 1,496′* MATERIAL *Steel and concrete* DESIGNER/ENGINEER *L. Wintergeist/ W. Tiedje* DATE COMPLETED *1977*

NAME *Köhlbrand Bridge* LOCATION *Elbe River, Hamburg, West Germany*

TYPE *Suspension* LENGTH *Main span 1,066'; overall length 2.2 miles*

MATERIAL *Steel and concrete* DESIGNER/ENGINEER *Rheinstahl AG*

DATE COMPLETED *1974*

NAME *Rhine Bridge* LOCATION *Rhine River, Neuwied, West Germany* TYPE *Suspension*
LENGTH *Main spans 771′ and 695′; overall length 1,591′* MATERIAL *Steel and concrete*
DESIGNER/ENGINEER *H. Homberg/E. Jux* DATE COMPLETED *1978*